Spondon's Power Stations
A century of change and innovation
1917 to 2017

Peter Heath and Alan Hunt

Edited by Anita Hayes and Ruth Hooper

Spondon Archive

Other books in this series

Spondon Clubs and Societies 1851-2011
The Malt Shovel, Spondon
Canoes to Commuters: transport in Spondon through the ages
Exploring Spondon: walks through Spondon's history
The Parish Church of St Werburgh's, Spondon
The Methodist Church, Spondon
Spondon School 1662 to1839
Spondon School 1839 to 1964
Spondon goes to the Pictures: a history of Sitwell Cinema
My Life in Spondon and other great places
Scouting in Spondon

Spondon Archive is the publishing name of Spondon Community Association and Spondon Historical Society. It was set up as a 'not for profit' partnership to record the history of Spondon.

ISBN 978 0 9930949-1-0

Printed by

©2017 Spondon Archive

Contents

Introduction .... Page 1

1 British Cellulose and Spondon's first Power Station 1916 to 1922 2
First Power Station 1916 to 1922 2
Derby Power Station 1893 to 1960 4

2 Spondon A 1922 to 1982 7
The National Grid 1926 9
Midland Counties Electricity Supply Company (MIDESCO) 9
The Generators and Turbines of Spondon A 10
Boilers and Steam Generation at Spondon A Power Station 12
Coal 15
Movement of coal by road 15
Movement of coal by rail 17
Tragic death of the Stationmaster at Spondon 21
Water 24
The beginning of the end for Spondon A 26

3 Workforce and aspects of employment 27
Bus Services 30
The Workers' Train 1920 to 1960 30
Murder at the Power Station 32

4 Spondon H Power Station 1959 to 2017 33
British Celanese events which paved the way for Spondon H 35
The layout of Spondon H 37
Commercial aspects of Spondon H 38
Winding Down 39

5 Spondon's Last Power Station and Derwent Cogen - 1995 to 2012 43
Gas turbines 43
Steam turbines 44
Fuel 44
Auxiliary boiler 46
Cooling water 46
Boiler feed water 46
Inlet air for gas turbines 46
Staffing levels 46
Cold start 46
Power generation 46
Final Closure 47

Acknowledgements 48

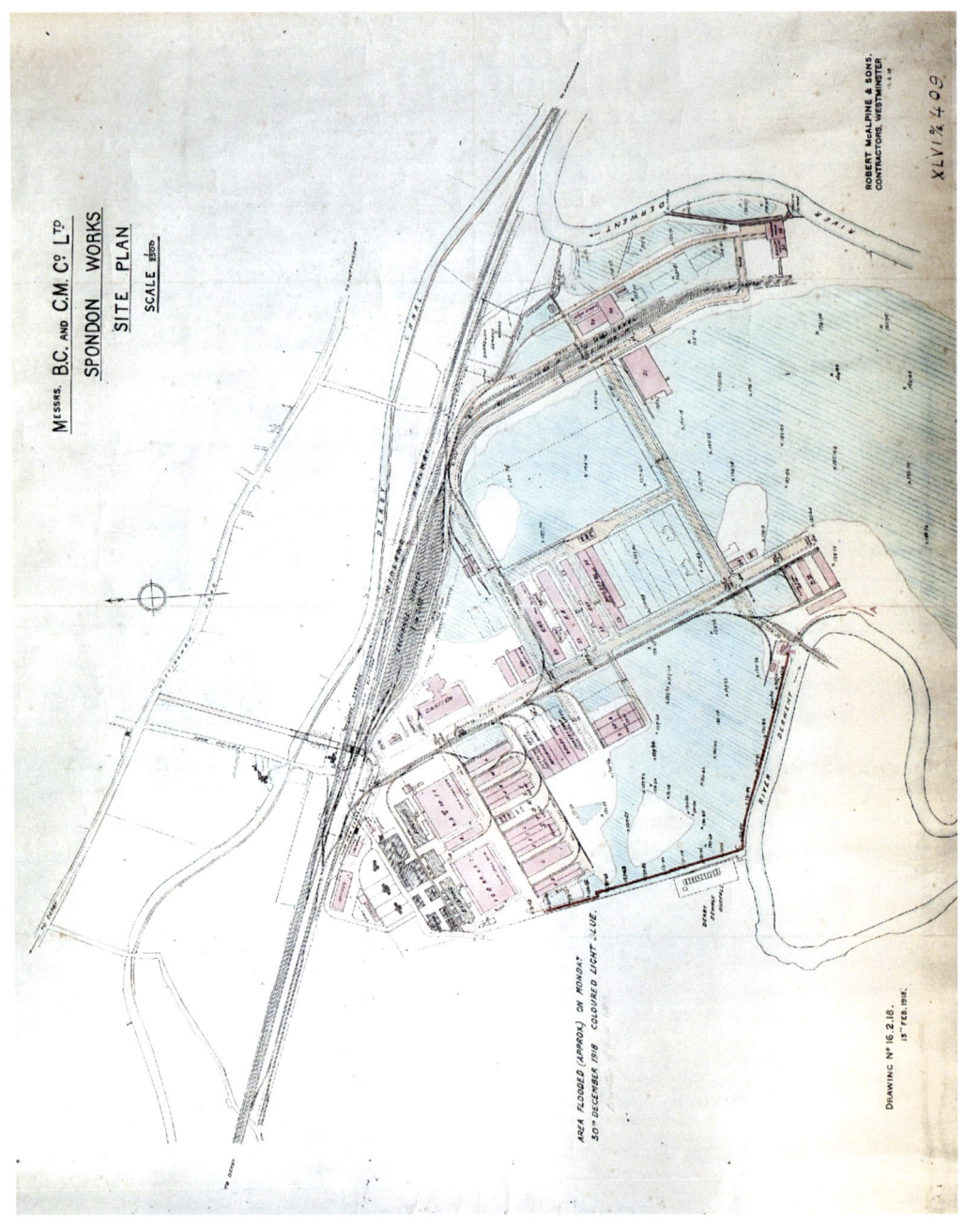

December 1918 McAlpine plan of site. The first Power Station was located at the eastern end of the site, next to the river.

See centrefold for plan of Spondon's Power Stations added to the April 1961 Plan for Spondon H.

Introduction

At 2pm on the 12 December 2012 Spondon's last electricity generator was switched off. This ended 95 years continuous production of high pressure steam and electricity from four separate power stations. At its peak Spondon supplied about five percent of the UK power needs, from the mid 1930s up to the early 1950s.

The first power station was built by the British Cellulose and Chemical Manufacturing Company between 1916 and 1918 and commenced supplying power to the factory around 1917, eventually removing the need for an earlier supply from Derby Power Station.

The Company contribution to the war effort ended in 1918 and they were faced with finding new markets for their products. In 1922 the Power Station was sold to the Derbyshire and Nottinghamshire Electric Power Company, who went on to develop it along commercial lines. In 1926 the Government set up the Central Electricity Board to design and build the National Grid with the eventual aim of linking together all of the UK power stations to be operated under one central control. This station became known as Spondon A in 1948. By the early 1950s Spondon A was being overtaken by the new Trent Valley stations and its end was in sight.

Likewise, the steam boilers of British Celanese, dating back to 1918, were well past their retirement age, so it was seen by the engineers of Celanese and Central Electricity Generating Board (CEGB) that there would be a mutual benefit in building a Combined Heat and Power Station. This was completed in 1959, known as Spondon H and rated to supply up to 200 tons of steam per hour to the Celanese site. As a by-product about 30 megawatt (MW) of power was also exported to the National Grid.

Spondon H was a very small part of CEGB. In 1982 it was sold back to the factory (now Courtaulds) who bought it and continued to run it using their own staff. By the early 1990s Spondon H was struggling to meet legislative requirements on dust and flue emissions and in its turn was now being seen as past its 'sell by' date.

Spondon's fourth and final Power Station entered service in 1995. It was a highly efficient state of the art Combined Cycle Gas Turbine station, and could reach full output from a cold start in only 45 minutes. It was profitable and had paid for itself, but closed when the factory closed in 2012.

By early 2018 it is likely that nothing will remain of any of Spondon's power stations or virtually any part of the factory buildings either; consequently we felt the need to write this story as a permanent record and tribute to the many hundreds of skilled and dedicated workers at every level who made it all possible.

Peter Heath, Site Services Electrical Engineer, worked for Celanese from the age of sixteen, and was responsible for all site mains power from 1970 until his retirement in 1997.

Alan Hunt served a full apprenticeship with CEGB and remained with them until 1969 before transferring his services to the factory site where he was employed on a wide variety of electrical projects until his retirement in 1999.

1 British Cellulose and Spondon's first Power Station 1916 to 1922

When the British Cellulose and Chemical Manufacturing Company built its works at Spondon the only suitable source of electrical power was Derby Power Station, located in Derby at the corner of Sowter Road and Silk Mill Lane, so a private on site power station had to be included in the works.

A number of major changes and events then took place in both Spondon and Derby and in order to see the whole picture it is necessary to take a wider view than trying to focus on the British Cellulose Power Station alone.

Derby Power Station had to adapt and grow in order to supply the British Cellulose plant at Spondon while its own power station was being built on a new 300 acre site. The Midland Railway Company had to supply and install a much extended freight handling system at Spondon. The work commenced in August 1916 and was basically completed by December 1918.

£5 million of Government backing was provided through the War Office for the erection of buildings and plant on both the Factory and Power Station sites. When World War 1 started, aircraft technology was in its infancy and payments were made to the Dreyfus brothers to continue to make a product called dope in Spondon. This was used to tighten and stiffen the fabric stretched over airframes and render them airtight and weatherproof. Dope consisted of cellulose acetate dissolved in acetone. Only three factories produced this product at the time, these were based in Germany, France and the Dreyfus factory in Switzerland.

As an example of how urgently this work was regarded by the Government they arranged that men of call-up age would be excused military service if they had the requisite skills to offer.

The first batch of cellulose acetate was produced in April 1917, after which production continued to increase in both quantity and quality.

First Power Station 1916 to 1922

British Cellulose and Spondon's first Power Station 1918

The Derby Telegraph of 15 June 1916 shows two pictures of the Company's station under construction. One shows the steelwork which looks complete and the other shows the station finished except that one of the two chimneys is only about two thirds erected.

This picture is from August 1918 and shows a wisp of smoke or steam emerging from the top of the one completed chimney so it is possible that they were not producing power at this time. However, there is firm evidence from a book called 'The Electric Furnace' written by John Norman Pring and published in London in 1921 that calcium carbide was first produced at Spondon in 1918. The only possible source of power for the electric arc furnaces at that time was the newly erected power station. It can therefore be said with a fair degree of certainty that the station was exporting power before the end of 1918. An advertisement of July 1919 states that Spondon was the largest producer of calcium carbide in Britain.

Turbine Hall 1917

A picture taken in 1917 shows the completed turbine hall with two generators, each about seven feet long and about the same height, with a well-dressed and hatted man standing on the non-drive end of the bed plate. Each generator was rated at 6MW and was direct coupled to a steam turbine with a speed of 1,500 rpm and the machine supplied 3 phase power at 6,600 volts.

The main purpose of Spondon's original Power Station was to provide power for the manufacture of calcium carbide in the electric arc furnaces erected in 31 Building. Calcium carbide was the route to the manufacture of acetic acid via acetylene gas. After 1916, and until the Power Station was commissioned, power for the site must have come from Derby, maybe until some point in 1919 when the first 6,600 volt power cables were laid between the Power Station and Numbers 1 and 2 Site Substations, later known Substation A and Substation B. 'A' provided power for the manufacture of Cellulose Acetate, 'B' fed the acetic acid production plants and a new Substation C was built to supply the yarn producing area known as Spinning.

Substation C was connected to 'B' by three 6,600 volt cables, so at this time all four major manufacturing areas of the site were connected to the Company's own power station as well as to Derby via the overhead line. Bearing in mind that the Company was still under an obligation to take a minimum of 1,000MW hours per month from Derby Power Station, the day to day management of the system would have required careful observation to ensure that Derby obtained its share of the load.

Derby Power Station 1893 to 1960

This Power Station is included as it played an essential part in the start-up of British Cellulose, and is part of the broader history of Derby.

A large scale map of Derby North published in 1899 indicates the station as 'Electric Light Works' and for most its life the station was owned and managed by Derby Corporation. It commenced generation in 1893 and was listed in February 1901 as having an output of 44,000 lampage units. Output was measured in this way because virtually all electricity produced was used for lighting and one lampage unit was equal to 8 candles, so the output of Derby Power Station would have replaced about 352,000 candles. The only lamps in use at that time had carbon filaments and their efficiency was a very low, 3 lumens of light for one Watt of power. One candle equals about 12.5 lumens, so the output of the station would have been about 1.4MW.

Demolition of Silk Mill House to make way for the 1908 extension to Derby Electric Power Station
Source: WW Winter 1907

Until 1908 it is a virtual certainty that the power produced would have been in the form of Direct Current (DC) using DC generators direct coupled to reciprocating steam engines. Some of this would have been for electric trams because they replaced horse drawn trams in Derby in 1904. DC power has to be used at its generation value, almost certainly at 460 volts which was then split across the 'neutral' conductor 230-0-230 for use by domestic and small business consumers.

1908 First steam turbine on the right and on the left is a rotary converter

On 2 February 1918, the Derby paper refers to these works, and reports that, acting upon a suggestion from the Ministry of Munitions, Derby Corporation approved a 6MW extension to Derby Power Station at a cost of £87,000 'in order to provide a further supply of power to the Company'. But circumstances were changing, and these works were not immediately carried out. They were completed by 1921 but went on to benefit the town of Derby as a whole because the slender overhead line already in place between Derby and Spondon would have struggled to handle 1.5MW, let alone anything like 6MW, and no evidence has been uncovered to indicate that a second line was ever installed.

An earlier source of civic pride along the river was the Derby Electric Power Station, even though it hid the historic Silk Mill and its chimneys vied with the ancient tower of the Cathedral. It was built in 1921. This party of official visitors is being shown around the wonders of power generation.

Source: WW Winter and Derby Telegraph Bygones

Peter Heath recalls being an apprentice electrician at British Celanese in the early 1950s and being taken into the 'Old Derby Sub' on the east side of 12 Extension Building located on the south side of Alfreton Road (see page 14). This was where the line from Derby terminated into an ancient cast iron 6,600 volt switchboard, probably no more than six feet in length and it definitely did not have a second supply; apart from anything else the building itself was only about ten feet in length.

The supply from Derby was carried on wooden poles for a distance of three miles. It was energised at 6,600 volts 3 phase and terminated on a pole at the back of what became the original Staff Canteen situated on the east side of 28 Building. From there it was routed underground by a 0.1 square inch cable into 'The Old Derby Sub'. Under the terms of the agreement with Derby Corporation the Company was limited to a supply of 3MW and had to pay for 1,000MW hours per month whether it was used or not, so the average load would have to be 1.37MW in order to avoid a penalty charge.

At this level the load current would be 150 Amperes and volt loss due to transmission about seven percent. This would have been an acceptable load and would work without placing undue stress on any of the components.

The overhead line from Derby is shown on an original Midland Railway drawing dated December 1921 and the chain dotted line indicating the route appears to follow the line from Spondon Sidings to Chaddesden Sidings. From there it would have been taken across the fields to the junction of Sowter Road and Silk Mill Lane where Derby Power Station was situated.

In 1901 there were 174 power stations in the UK, including what is now Eire, and they produced power in a wide range of voltages and frequencies as well as a large amount of direct current. Frequencies ranged from $16^2/_3$ (London Underground) up to 90 cycles per second in one area, making it impossible to interconnect stations so that they could support one another when needed. This meant that each station stood alone and they all made their own decision as to what supply to offer to their consumers. This was the situation in 1918 and would remain so until 1926 when the National Grid came along.

In 1908 the Corporation installed Derby's first turbine generator which would have produced alternating current, probably with a rating around 3MW. The next news came from the Derby Daily Telegraph dated 3 October 1917, when Alderman Wilkins, chairman of the Corporation Electricity Committee, stated that the British Cellulose Company was arranging to make its own electricity. He went on to propose that the sum of £725 be approved for the essential repairs to 'the armature of a turbine', these being required because he said it had already done ten years hard work. Presumably, therefore, this would have been the armature from the 1908 machine and that by 'armature' he meant the rotor which is the central rotating part of an alternating current generator. Finally Alderman Wilkins said that there would be an interim period while parts were awaited and this would lead to a power shortage which could last until January 1918. Was this due to the expectation that supplies to the Company would adversely affect the availability of power for Derby? It fits the overall picture at the time and was probably the case.

Prior to the Derby paper report of 3 October 1917 there was an earlier paper dated 2 August 1917 which reported that 'certain outstanding questions had to be settled with the company' before providing plant 'as may be necessary for affording a supply'.

The Mayor resolved that they should apply to the Local Government Board for sanction to borrow £90,000 for the purposes mentioned. Alderman Wilkins then added that this sum could now be reduced to £20,000 due to a likely arrangement with the Government. The resolution was carried.

A later Derby paper report on 29 June 1918 refers to the final Supply Agreement between the Company and Derby Corporation.

Derby Power Station continued to generate for about a further 40 years but the passing of the Clean Air Act of 1956 signalled its end and it stopped producing about 1960.

Derby Power Station 1939

2 Spondon A 1922 to 1982

After the end of the First World War the market for cellulose acetate virtually disappeared and the Company had to look for other outlets or products in order to survive.

This was a critical period and the Company came near to closure, so they had to look for aspects of the business which could be sold in order to raise capital. An obvious choice was the Power Station; this was sold in 1922 and became known as the 'Derbyshire and Nottinghamshire Electric Power Station'.

It wouldn't be long before the Derby & Notts Electric Power Company would be looking for additional customers, but to do that they would need extra capacity in order to ensure a secure supply; in 1923 a further generator rated at 12MW was installed and this ran at the now standard speed of 3000rpm. The first two 6MW machines were referred to as Numbers 1 and 2 and the new machine was given the title Number 1A.

Spondon A was Spondon's second Power Station and was built around the first one, so that until 1929 you would still have been able to see the original building.

Spondon A Power Station 1946

Power stations have to have a 'black start' capability because if they stop and go cold restarting them is a problem, but help would have been at hand from the factory's 6,600 volt overhead line from Derby. This had a connection to Number 1 Substation which was linked directly to the Station's 6,600 volt switchboard, so up to around 1.5MW could have been fed back, more than enough for a 'black start'.

In the Derby Daily Telegraph of 3 May 1922, under the headline of 'Current to Big Works', Alderman Wilkins, Chairman of Derby Corporation Electricity Committee, referred to the need for a 'rearrangement' of the tariff because of the change of ownership of Spondon Power Station. On 29 July 1922, under the headline 'Agreement varied', details were given of the 'rearrangement'. The previous minimum average demand was raised from 1000MW hours per month to an average minimum load of 1.5MW; this to be taken equally by the Company and the Power Station. These changes were necessary in order to ensure that Derby Corporation received a satisfactory income from the large investments they had already made.

By July 1922 it is a virtual certainty that the Company load already exceeded 1.5 MW so the change would have been a simple accountancy exercise. The real gain was that a permanent power link was now in place between Derby and Spondon giving operational benefits to both parties. The agreement was for six years from 22 July 1922 and lasted until the National Grid came along in 1926.

Battery/Generator Control Panel 1932

These panels were typical of both Power Station and Factory. They were known as the Frankenstein panels because of their likeness to the panels in the Boris Karloff film of that name made in 1931. The Spondon panels were made in 1932; the larger one of the two is restored and working in the Papplewick Museum pumping station near Nottingham.

The National Grid 1926

The Government set up the Central Electricity Board in 1926 and they were given the task of designing and building the National Grid with the aim of linking together all of the UK power stations on a standard frequency of 50 cycles per second.

Spondon's National Grid connections from 1926 at the rear of the picture, photographed 2016

Peter Heath recalls having regular meetings during his early years on the power system with a British Thomson Houston sales engineer who told him that he was aware of his Company's involvement in the supply of Derby & Notts Power Company first 33,000 volt switchboard sometime around 1924. This was a huge weatherproof outdoor metal-clad arrangement which had duplicate bus bars with the very unusual arrangement of having two oil filled circuit breakers for each circuit connected. Each one comprised three separate single phase tanks, thus providing flexibility and a high degree of security for the supplies.

This switchboard became the main distribution point for power export from Derby & Notts and was linked to the National Grid through two transformers which Peter Heath recalls as being rated at about 100MW each.

Midland Counties Electricity Supply Company (MIDESCO)

Derby & Notts became part of Midland Counties Electricity Supply Company in 1922 with offices in Ilkeston. There was a reference in newspapers of the day to a new Power Station starting up at Spondon under the heading 'Super Power Stations'. MIDESCO was first registered around 1910 under the name of the Tramways Light and Power Company Ltd. for the purpose of owning and working the Derby & Notts, Leicester & Warwick and the Leamington & Warwick Power Companies; as well as the Notts & Derby Tramways Company. In 1919 their stations were at Hinckley (Leicestershire), Avon Bridge (Warwickshire) and Ilkeston (Derbyshire).

This ties in well with the fact that in 1924 double circuit 33,000 volt overhead lines were installed linking Spondon with Hinckley, Ilkeston and Avon Bridge Power Stations. This enabled these stations to be worked in parallel with each other and weekend, night and holiday loads were supplied from Spondon only. This arrangement lasted until 1941 when Avon Bridge received a 132,000 volts connection from the National Grid.

The Directors realised the importance of standardising their system at 50 cycles per second and in one weekend during January 1923 they changed all of the system from 25 to 50 cycles per second, it was said without any interruption of supply. Ilkeston Power Station then closed for good and the town was re-fed from Spondon. Ilkeston ran from 1903 to 1923.

Avon was known as Avon Bridge and was situated midway between Leamington and Warwick in a district known as Emscote, about 43 miles from Spondon. This must have been about the limit for sensible transmission at 33,000 volts because at times of light load the capacitance of the line would have had a significant effect on voltage regulation, i.e. the voltage at Avon would have been higher than the voltage leaving Spondon. It probably required careful supervision at these times and this arrangement continued until the arrival of a 132,000 volts connection from the National Grid in 1941.

Under the 1926 Act the generating stations of MIDESCO became 'selected stations' and had to be operated under the direct instructions of the Central Electricity Board.

The Generators and Turbines of Spondon A

Generator Size in MW	Year Installed	Original Number	Number After 1941	Year Scrapped
6	1917	1	-------------	1941
6	1917	2	-------------	1941
12	1923	1A	3	Oct 1973
15	1926	2A	4	Mar 1974
25	1929	3	1	1982/83
30	1930	4	2	1982/83
30	1935	-------------	5	1982/83
30	1939	-------------	6	1982/83
30	1944	-------------	7	1982/83

This record shows the rapid growth in capacity of the Derby & Notts Station after acquisition in 1922, and by 1929 a further 52MW had been added to the original 12MW, an increase of 400 percent in only seven years.

15MW turbo generator 1926

A very serious incident occurred during the early 1960s. At around 4 or 5am instructions were received from the Grid Control Centre in Birmingham to prepare Number 6 30MW machine for service to meet the usual morning load peak as the nation began its working day. Turbines of this size have to go through a rigorously controlled procedure to ensure that the fixed and moving parts are 'warmed up' in unison, because internal clearances are very small and uneven heating can cause them to expand at different rates resulting in collisions. Significant damage can be caused in this way.

As the machine accelerates it passes through resonant speeds which can result in serious vibrations. So the work of the turbine driver is of a very responsible nature and requires a good deal of training.

Alan Hunt was at work in Spondon A on that morning and clearly recalls the sound of a loud explosion coming from the northern end of the turbine hall. He was on the scene shortly afterwards and describes seeing the low pressure end of Number 6 Machine with a large hole blown through its eastern side.

The time was almost 8am and preparations for putting the machine on load were in their final stages with the turbine running at something like 80 percent of service speed of 3,000 rpm. At this point the low pressure wheel, the machine's largest, disintegrated and passed through the thick steel casing of the turbine, ejecting large pieces of metal which shot across the turbine hall and struck a two inch thick slate panel upon which were mounted various meters and relays. Behind the panel was the east wall of the hall and this was left undamaged.

A lengthy enquiry followed involving the CEGB, the British Thomson Houston Company who made the turbine and most probably the Factory Inspectorate but the final conclusion and result remains unknown.

No one was injured in this incident and the turbine was eventually either repaired or replaced by a new machine.

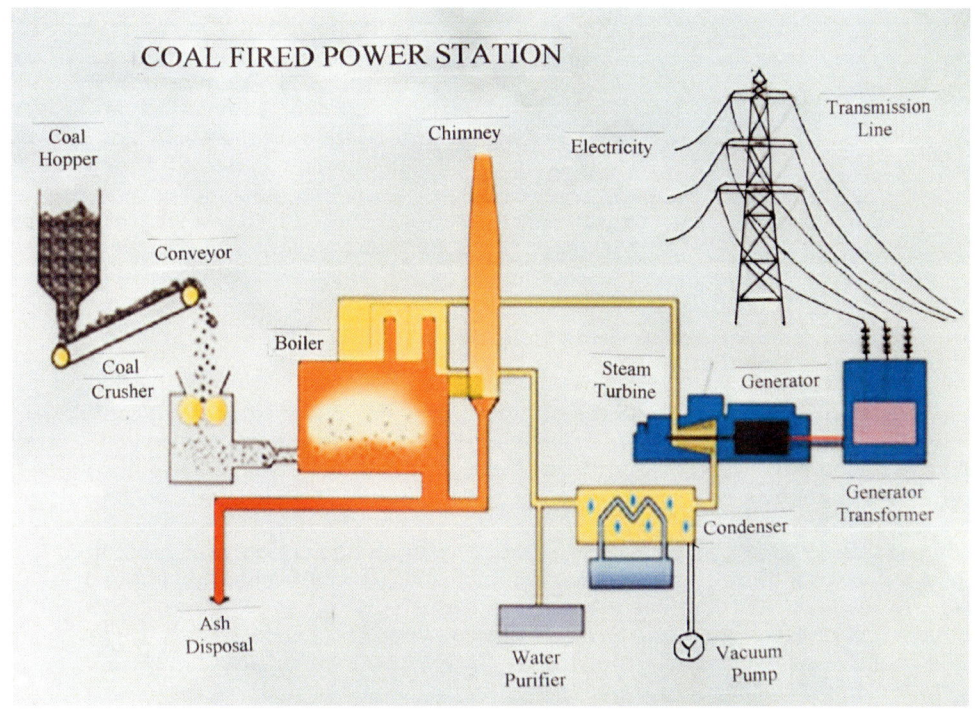

Traditional coal fired power station

Boilers and Steam Generation at Spondon A Power Station

All of the steam produced at Spondon A was used to produce electricity for the National Grid and none of it went into the factory site. The factory produced all of its own steam in Number 86 Boiler House until the completion of Spondon H Power Station in 1959.

Number 2 Boiler House Firing Floor in Spondon A, 1929

The output of Spondon A grew to 172MW between the years 1923 and 1944 and this increase was accompanied by a commensurate rise in the steam requirement.

The original Number 1 Boiler House had two chimneys. Both were connected to a bank of three boilers, each of which was rated at 11 tons of steam per hour. Number 2 Boiler House was added on the north side of Number 1 to match the steam needs of the Station.

By 1929 there was an expectation that the station would continue to grow. When the 25MW generator arrived, Number 3 Boiler House was added on the north side of Number 2, and this had eight boilers, each boiler with its own chimney. Total steam capacity was about 500 tons per hour. This could support 100MW of generation.

While these works were going on the original Number 1 Boiler House was completely demolished and in its place a new Boiler House, also called Number 1, was built. The boilers installed here were the largest of the Spondon A boilers and again each one had its own chimney.

So by the late 1930s or early 1940s the Power Station had a total of 22 coal fired boilers and a total of 22 chimneys, each one discharging into the atmosphere at a relatively low level. With an easterly wind these exhaust gases fell directly onto the southern half of the factory and Peter Heath has vivid recollections from the early 1950s of their positively sulphurous nature, to the point where breathing itself was unpleasant and difficult.

The great London 'smog' of the early 1950s focussed national attention on this problem and led to the closure of Battersea Power Station and many others. Spondon's response to this situation was to erect a single large chimney 350 feet high with an inside diameter at the top of 14 feet; it served the largest six boilers located in Number 1 Boiler House. To do this the flueing arrangements had to be rebuilt so that the chimney was connected to the boilers in two groups of three. All the other boilers continued in service with their original chimneys.

Spondon Power Station early 1950s
Turbine hall in foreground, next to River Derwent
Boiler house at rear
Large building on the right is 29 Building machine shop belonging to Celanese

The final boiler arrangement at Spondon A was as follows.

No. 1 Boiler House had six boilers numbered 1 to 6, each with a nominal rating of 71.4 tons/hour.
No. 2 Boiler House had eight boilers numbered 7 to 14, each with a nominal rating of 17.9 tons/hour.
No. 3 Boiler House had eight boilers numbered 15 to 22 each with a nominal rating of 35.7 tons/hour.

Number 2 Boiler House was scrapped in 1974 and this coincided with the scrapping of Numbers 3 and 4 turbines.

At this point it is worth commenting on the Factory road-naming system. The road names progressed alphabetically from the main gate and on the left hand side they were Buxton, Derby, Filey, Heanor, Ilkeston, Kedleston, London and Nottingham. Those on the right were Alfreton, Cambridge, Eastwood, Grantham, Jarrow, Locko, Matlock and Ollerton; the first letter of each name was used for side streets. So Eastwood led to Edale and Elvaston Streets; side roads were all Streets.

Although Celanese Road was mostly outside the Factory and Power Station fence, this is the only remaining example of the road signs used

The Spondon Factory site always had a steam pressure of about 160 pounds per square inch (psi). The original supply was from two boilers in 13 Building at the west end of Alfreton Road and another two in 23/B at the west end of 23/4 Building, at the west end of Kedleston Road. The original British Cellulose Power Station was built at the same time and had six boilers, none of which was connected to the Factory site so it may be reasonable to assume that all ten boilers were ordered together for a common nominal pressure of 250psi and this has been confirmed by David Howard, who went on to become the Manager of Spondon H Power Station in 1983.

This pressure of 250psi fits in well with known power station pressures of the period 1917 up to the early 1920s, so it is most likely that Spondon's first Power Station operated at this pressure.

Power station records show that pressures reached about 650psi nationally by about the mid 1930s and a record of 1920 gives North Tees a pressure of 450psi. Spondon's output was only 12MW so the lower pressure looks about right. For these conditions the 1917 Power Station would have burned about 1,200 tons of coal per week and this was delivered by rail.

When Number 2 Boiler House was built during the mid to late 1920s a higher steam pressure than the earlier 250psi would have been adopted but its exact value is unknown. The most common pressure in the UK after the late 1920s was 625psi and after that the next step upwards was to pressures in excess of 1,000psi but there was never a general rule. A notable exceptional was the Ford Motor Company Power Station at Dagenham; it ran at 1,250psi in 1930. Brimsdown (Enfield) ran at 2,000psi in 1938, but these were exceptions.

After the original Boiler House was demolished to make way for the new Number 1, the two 6MW machines would still require a 250psi supply and this was obtained through pressure reducing valves, connected to the new higher pressure, which is most likely to have been around 625psi.

Coal

The coal for boilers with chain grate stokers has to have a fairly high ash content in order to protect the moving grate from the intense heat at the point of combustion. A typical ash depth for satisfactory operation would be about four or five inches and this requires an ash content of about eleven percent. The calorific value of coal such as this is about 11,000 British Thermal Units per pound and a reasonable estimate for the thermal efficiency of Spondon A in the early 1940s is no more than 20 percent. As an example, the average efficiency of all UK power stations in 1951 was only 21.54 percent so 20 percent could be a little optimistic for Spondon with its 22 individual relatively small boilers. Relating these figures together is a simple calculation which produces a coal consumption of about 20,000 tons per week for an output of 170MW.

Until about 1945 coal was delivered exclusively by rail and tipped from the wagons into low level hoppers. From here it was lifted by bucket elevators to the highest level above the boilers and then moved on conveyor belts to positions above the boilers where it was diverted onto the gravity fed hoppers so that it fell directly onto the moving grate below. At this point it was roughly spread across the grate by a moving arm before passing below a levelling bar which set the depth of coal across the width of the grate.

The grate was driven by a small fixed speed electric motor connected to a variable ratio gearbox and this provided a range of speeds so that the rate of firing could be set and accurately held.

Below the grate on the floor below were the forced draught (FD) fans and the flue gases were then pulled out of the boiler casing by the induced draught (ID) fans. Air control was by adjustable dampers and the FD and ID fans were balanced so that the pressure inside the boiler casing was always slightly negative. In this way virtually no injurious fumes escaped into the manned area of the boiler house.

Movement of coal by road

By the end of World War Two, coal deliveries to the station would have been about 20,000 tons per week so it should come as no surprise that around 1945 Derby & Notts decided to build a roadway between Elvaston Lane, Borrowash (now called Station Road) and the Power Station so that coal could also be delivered by road; a major advantage of this scheme was that boiler ash could be taken away.

The roadway came to be known as Power Station Lane, it was about 1,500 yards long and made two crossings of the River Derwent, the second one by means of a Bailey bridge at the point of entry into the Power Station.

**Bailey Bridge 1960
On the left is the Stores
On the right are the Main Office Block, and Works and Staff Canteens**

Again, Alan Hunt remembers that tipper lorries came into the Power Station throughout the day in a continuous procession and generally tipped their load directly into the coal unloading hoppers. The same vehicles would then be used to take away boiler ash, a most useful function since removal by rail would not have been a practical proposition. The ash was used for landfill, hardcore for building works or as grit for use on some roads during harsh winter conditions.

Perhaps Derby & Notts could see another advantage in having two delivery systems for a material that was essential for their operation: in the event of a pit or rail strike they would gain some flexibility of operation.

Deliveries would have been at the rate of something like 5000 tons per week brought in by heavy tipper lorries and a photograph taken in February 1972 shows two of them nearing the end of their journey, fully laden with coal. These were heavy vehicles with eight wheels at the rear and four at the front and each would have carried around 20 tons of coal. On the basis of 5,000 tons per week this would equate to about 34 truckloads per day, including weekends.

Movement of coal down Power Station Lane 1972

It is believed that coal delivered by road came from Stanley Colliery or other ones nearby, as there were several in the area. The route was via Dale Road, down Moor Lane to Ockbrook then along Victoria Avenue, followed by a short run down Elvaston Lane before turning right onto Power Station Lane.

Movement of coal by rail

**The first line to the site for the delivery of coal, April 1918.
On the top left is the platform for the workers' train.**

By the mid 1940s coal deliveries to the Power Station would have been at the rate of about 20,000 tons per week. The Factory also required large quantities of coal for steam-raising as well as cellulose in the form of wood pulp or raw cotton. The other need was for acetic acid, or something to make it from, such as naphtha.

The quantities involved were very large so it is not surprising that from the 1920s onwards Spondon's rail connections to the British Celanese/CEGB sidings developed as a major supply hub for all bulk deliveries. There were separate branch lines for the Factory site and the Power Station had its own coal sidings on the north side of Eastern Avenue. Coal was brought in using the old style wooden wagons which held about 20 tons each and a train load typically consisted of 20 or maybe 25 wagons so deliveries could have been at the rate of seven trains per day, seven days a week.

These trains would have left the main line to be diverted into the LMS sidings outside the Celanese Works before being pushed into the Power Station sidings along Eastern Avenue. From here they had to be pulled again by the Station's battery powered locos into the Power Station site for unloading. The empties then had to be reassembled and shunted back onto Eastern Avenue before being returned to the main line so it was a full time job.

Spondon Station 1924 showing first footbridge

Both sites had their own internal rail systems; the Factory had tracks along every main thoroughfare and employed a number of 0-4-0 saddle tank locomotives between 1918 and the mid 1960s, after which they were gradually replaced by diesel hydraulics known as DH1, DL2 and DL3. The steam shunters had names and the most memorable of them were Hector, Henry, Victory, George and Frederick. Frederick was unique. It was built by Barclay & Co. of Kilmarnock in 1867 and rebuilt by them in 1908 then resold to Stewart & Lloyds who sold it again to Celanese at an unknown date. It was finally scrapped in 1961.

The Power Station had three battery-powered locos for shunting duties and they were known as Numbers 1, 2 and 3. These were built by English Electric between the mid 1920s and 1946; each one having a gross weight of 20 tons with a 35 horse power motor mounted on each axle. They were fitted with two banks of lead acid traction batteries fore and aft of the steeple cab, nominally 200 volts. Battery charging was provided overnight in a loco shed.

1946 saw the purchase of the last battery shunter and this was also around the time that Power Station Lane opened, so after that each loco probably had to move about 5,000 tons of coal per week during something like a 50 hour working week, as they did not operate during the hours of darkness. These factors add up to the movement of about 100 tons per hour per shunter and David Howard, who managed Spondon H station recalls that one loco could pull about four or five loaded coal wagons per trip. So each one would have to make about seven round trips per shift on the basis of a seven day week, something like one hour per trip which makes sense when one considers the amount of work involved in unloading, reassembling the empties and returning them to the sidings.

Both Spondon and Avon Bridge Power Stations had extensive networks of two foot narrow gauge tracks and these were used for the transport of ash from the boilers. The locos for this duty were once again battery powered.

2 foot gauge electric loco
Copyright Ashover Light Railway Society

Standard gauge electric loco
Spondon A, Number 3 Boiler House in background

Returning to the subject of Spondon's electric shunters, David Howard remembers that these were sold shortly after the acquisition of British Celanese by Courtaulds, at the time when the Company purchased Spondon H from the CEGB. This was because the Company's own Diesel shunters were available for this duty and they could easily pull much heavier loads than the electrics.

Number 2 Electric Shunter went to the Foxfield Railway at Blyth Bridge, Staffordshire where it remains, but requires new batteries. Number 1 went to the ill-fated Transperience Museum (a short-lived museum of passenger transport located at Low Moor, in the south of Bradford in West Yorkshire, Northern England. It opened in July 1995, but closed only two years later, in October 1997). It went from there to the Electric Railway Museum at Baginton, Coventry. Number 3 was scrapped and because, as David Howard recalled, it had been damaged in a fire to the point where restoring it was not a practical proposition.

Alan Hunt has an original large scale LMS drawing dated 1920 showing the proposed growth and development of Spondon's Factory and Power Station sidings to cope with expected future demand. Principal amongst its features are the storage and handling sidings along the north side of Eastern Avenue. The intention was for these to be 12 tracks wide but plans by Celanese to erect 119 Building and others on the south side of Eastern Avenue reduced this number to four and by 1979 this had become three.

Coal would always be the major component in the freight mix brought to the sidings and by the 1950s something not much short of 20,000 tons per week would be needed by Spondon A and the Factory, in addition to road deliveries of about 5,000 tons per week from local collieries to Spondon A Power Station.

During the winter of 1965 fifty one freight trains per day either used or passed through Spondon and most of these started from Chaddesden Yard bound for the Nottingham area but others originated elsewhere, Carlisle to Leicester and Manchester to London being two examples.

Two coal trains per day linked CEGB sidings to Bestwood Park, North Nottingham, and a third went from Celanese sidings to Immingham, carrying imported coal back to Spondon.

A typical train would comprise 35 to 40 twenty-one ton hopper wagons, probably carrying about 750 tons and on the known basis of two trains per day this would have been about 7,500 tons per week, plus the 5,000 tons per week brought from Stanley Pit by road. Certainly during the winter months this would have been barely sufficient and it is known that a third source was from Denby Colliery which had a single line connection to the Derby line at Little Eaton and that coal for Spondon was purchased from here.

Coal trains for Spondon were generally hauled by LMS locomotives Class 4F, Black 5's and later the more powerful 8F and 9F were used. From time to time the very powerful Garratts would appear on the scene but they were not generally favoured, mainly on account of high fuel costs. A photograph taken from the bridge has been seen showing a Black 5 hauling a train of about 25 loaded coal wagons and this loco is on record as having a maximum tractive effort of about 25,000 lbs, so perhaps this could have been a general rule of thumb that a force of 1,000 lbs was sufficient to pull a standard 20 ton wagon.

Tragic death of the Stationmaster at Spondon

Henry Ernest Haines

As a conclusion to the railway part of the story it is worth recording some details of the sudden and tragic death of the Stationmaster, Henry Ernest Haines, who lived with his wife and children in what was known as the Station House at the east end of the up line platform. He was killed while sitting at his desk in the station building on Wednesday 13 August 1924. Under the heading of SPONDON SMASH the Derby Daily Telegraph of 15 August 1924 carried a full report of the inquest into Mr Haines' death which occurred when a lorry, heavily laden with nine tons of acetone, failed to stop in the distance between the brow of the canal bridge and the closed level crossing gates.

The vehicle was still travelling at between 4 and 5 mph as it drew close to the railway bridge at a time when a train could be seen approaching the crossing. In a last ditch attempt to avert disaster the driver swerved hard to the left hoping to come to rest in the space between the Station House and the side of the bridge but he failed to do so.

The west wall of the Station Building was struck by the 17 ton mass of the lorry at a point just to the north of the first window and it went on to embed over half of its length into the Station Master's office where Mr Haines was sitting at his desk in the south west corner. The whole corner of the building was demolished. leaving a space of 9 feet by 10 feet in ruins and the first floor collapsed onto the seated Mr Haines, killing him instantly.

Evidence from the owner was that they had 20 lorries and this one had received a brake check on Monday, just two days before its departure for Spondon and that no faults had been found. This statement was corroborated by a Borrowash motor engineer who confirmed that the brakes appeared to be in order apart from a bent rod at the front but this could have been caused in the collision with the building.

Summing up at the inquest the Coroner said there was no doubt that the driver had been taken by surprise when the crossing gates first appeared but he had done his best to halt the vehicle and there seemed to be no evidence of culpability or gross negligence on the part of the driver. The jury returned a verdict of accidental death and recommended that the bridge be looked at and that a proper road be made.

Evidence of the building repairs carried out on the Station Building after the accident are clearly visible from Station Road by viewing the ground floor window in the west wall and following the line of brickwork as it curves up and over the window frame, easily picked out because the newer bricks are of a slightly different colour from the 1839 originals.

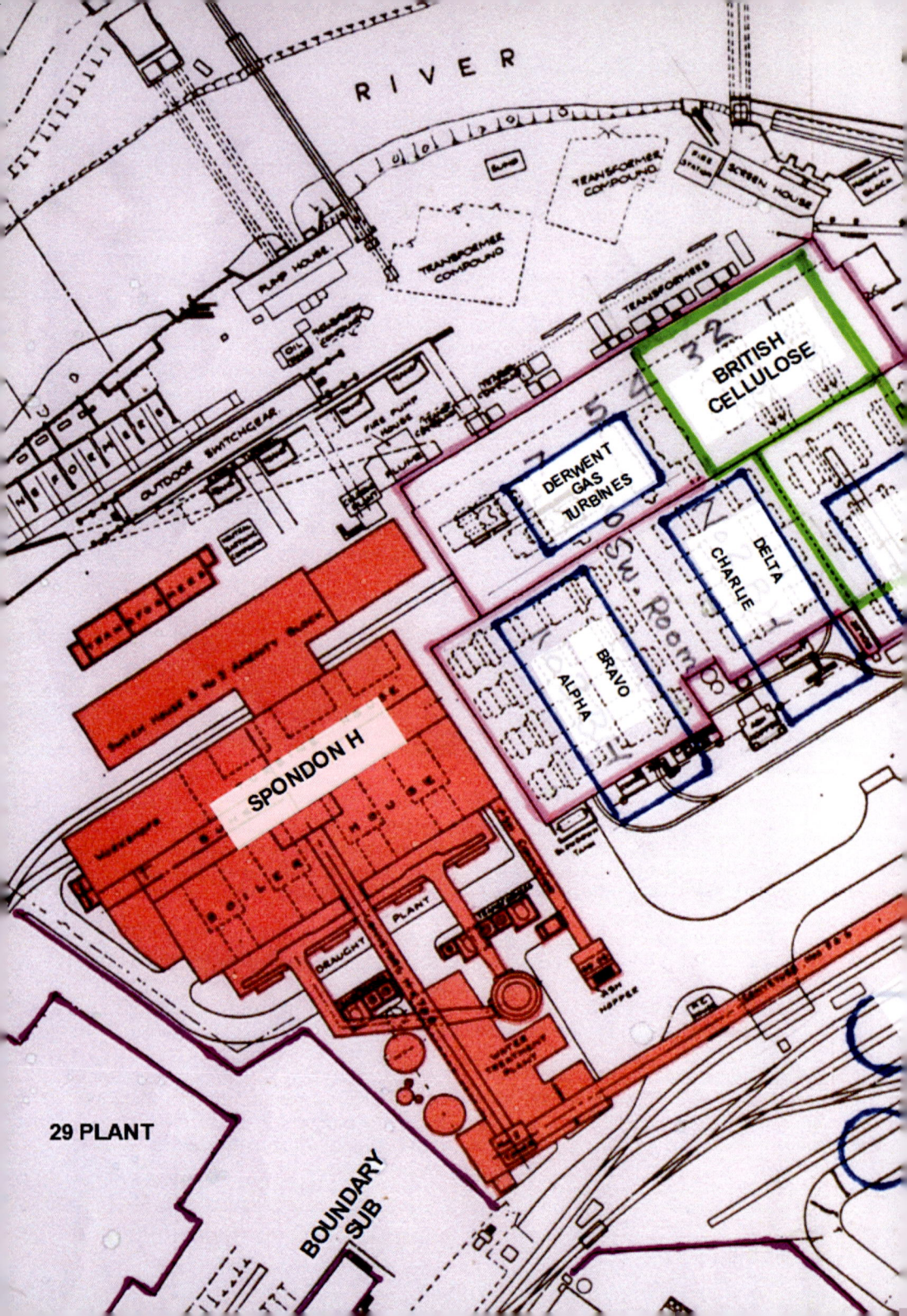

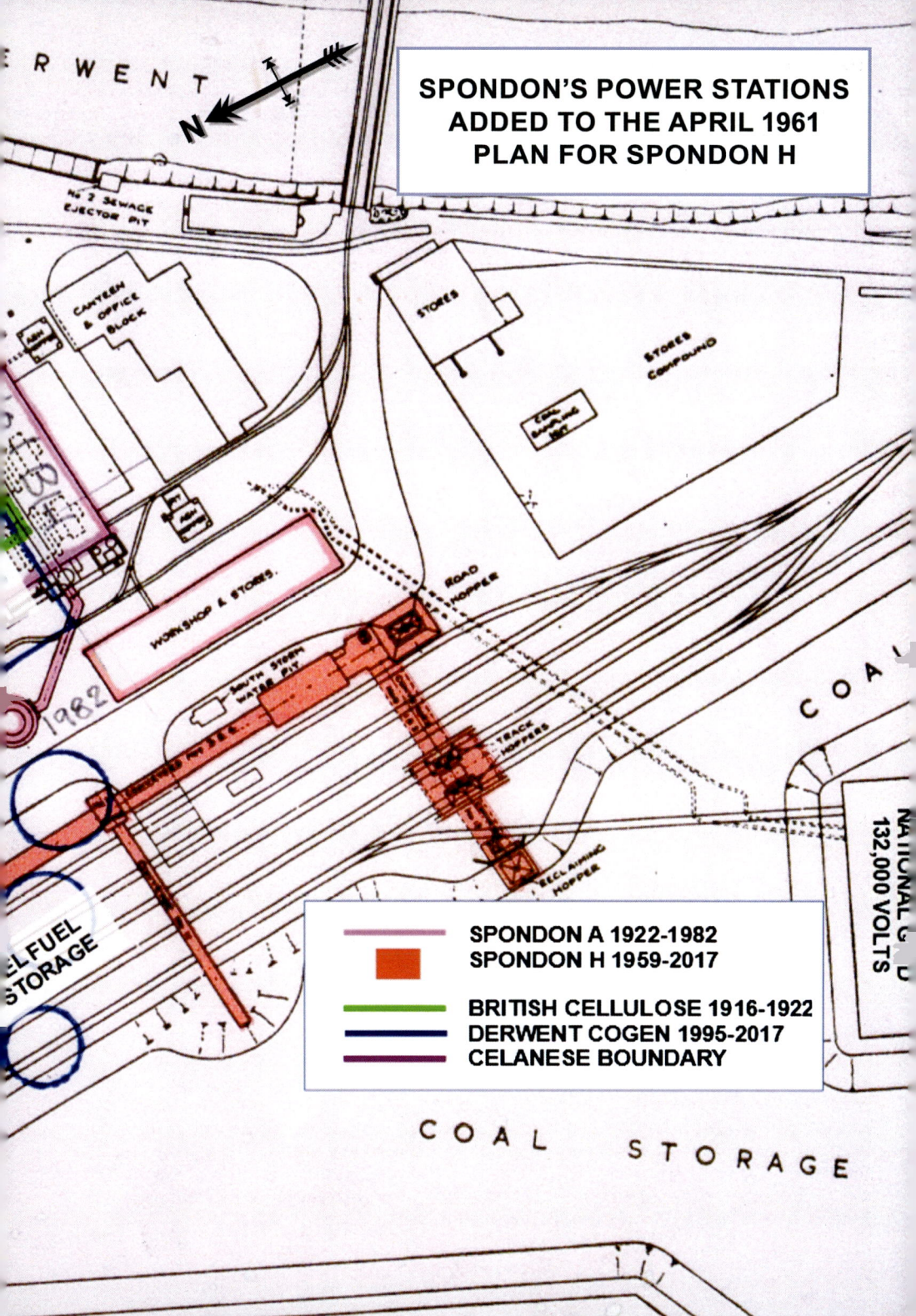

Water

From the earliest days the Company Directors had foreseen that the Factory would prosper and grow and large quantities of water would be needed by the Power Station and Factory site for steam raising and cooling as well as many other uses. Each had its own intake from the River Derwent. Two large interconnected lakes were scooped out over a period of about twenty years or so and they joined the River Derwent to the Lake Pump House intake. There was a dual purpose in these works; the first of these was to provide huge quantities gravel which was used to make concrete for all the new buildings that would be erected on site over the next three decades.

The second purpose, an important one, was to provide settlement beds for the silt and sludge which entered the lakes from the river. This material could not be allowed into the Factory because it would clog up the pipes and pumps very quickly.
So silt and sludge continually built up on the lake beds and every six to eight years it was necessary for them to be dredged and all of this material was then dumped on the south side of the river. This was not a problem as all of the land was owned by the Company and was tenanted by local farmers. About six months was allowed for drying and settlement and then the mounds were levelled and the land handed back to the farmers, who would speak highly of the fertile qualities of this material.

Lake Pump House. Original building of 1928 on the left, containing six pumps. The building on the right was the 1948 extension with two more pumps.

The River Pump House in the Factory was equipped with six 300 horse power pumps which each delivered about 350,000 gallons per hour at a nominal pressure of 40psi. In 1928 six more pumps were added at the Lake Pump House each rated at 465 horse power. Finally in 1948 two more were fitted so that the Factory water pumps totalled 5,500 horse power and were able to deliver a typical volume of about 100 million gallons per day into the site. Maximum summertime delivery has been recorded at about 120 million gallons per day but this was unusual.

There was also a Filter House, Building 47A, next door to the River Pump House and this provided about one million gallons per day of clean filtered water for use in the Factory. This water was used for product washing throughout the site and a major use was in the Dye House, Building 18, where clean sediment-free water was essential to the dyeing processes. Some processes demanded softened water and this was also provided but in relatively small quantities.

The return of cooling water to the river had to be watched carefully, especially during the summer, as the maximum permitted return temperature was set at 28°C. Above this temperature it would have been injurious to wild life, especially to fish.

Control of river level

The 1918 Power Station had six boilers each rated at about 11 tons of steam per hour and when all of this had passed through the turbines it had to be condensed back into water for return to the boilers and re-use. This was the purpose of the cooling water taken from the River Derwent and it had to absorb about 164 million British Thermal Units (BThU) per hour (48MW) before being returned to the river. Obviously, this amount of heat energy could not go into the river so the station must have had cooling towers, but they were low level structures probably made of wood.

By the 1940s the Station, now known as Spondon A, had an output of 172MW so by that time it had acquired four large conventional concrete cooling towers, two being built before World War Two and the others after. Each of these had its own river water pump which took water from the condensers and pushed it up into the cooling towers from where it was sprayed down to cool and surrender its heat energy to the atmosphere.

This was followed by a system of controlled diversions so that river water could go back into the condensers or some or all of it could be returned to the river. All of this was done manually to ensure that the temperature between the off take and return did not exceed about 4°C, with an upper limit of about 29°C.

The amount of energy lost in this way was huge, even the two 6MW generators in the original Station sent about 150 million BThUs per hour (48MW) into the sky and all of this energy came from coal which had to be paid for.

There was a noticeable effect of the warm water in the river in that it caused some wildlife to thrive. Alan Hunt recounted how large coarse fish including pike abounded where the warm water discharge took place and there were always swans' nests in the reeds. Swans out with their cygnets would return to the nest with one missing and it was commonly thought that it had been taken by one of the large pike which could be seen in the water.

Now, in 2017, the last of Spondon's power stations has been demolished and the River Derwent has returned to what it was a century ago; a reedy slow moving stream meandering its way to Shardlow where it flows into the River Trent. Peter Heath recalls that at the time of his retirement in 1997 the area around the lakes was known as The Nature Reserve and was open to employees, but there was never public access. There were painted signboards showing the general layout. The water in the lakes does not remain stagnant as there are two 'run offs' at the head of the second lake which return water to the river and that the flow is more than enough to keep the water fresh.

The beginning of the end for Spondon A
In 1945 the average electrical load of the UK was 3,450 MW and in 2015 it was a little over 44,000MW and from these two fixed points it is easy to calculate the average annual increase. The figure is near enough three percent per annum, so between 1940 and 1950 the respective loads would have been in the region of 2,980MW and 4,000MW. In 1940 Spondon A could produce 140MW; with the addition of another 30MW machine in 1944 output rose to 170MW. So over this period Spondon supplied about five percent of the UK requirement, commensurate with Ratcliffe on Soar today which has a continuous output of 2,000MW. The peak demand in the UK for the winter of 2014/2015 was 44,000MW.

By 1950 Spondon A was outdated and outpaced by a rising national load; as the 1950s progressed we saw the large building programme of a new generation of coal fired stations as well as the advent of nuclear power. Castle Donington opened in 1956 and had ten 100MW machines so the 'writing was on the wall' for Spondon A. From the early 1950s to the early 1960s the plant only ran close to capacity during the winter months. Alan Hunt again recalls the period when only two machines were on load at certain times of the year, during the summer period there were none.

The station finally went cold in 1980. The high concrete chimney came down in 1982.

3 Workforce and aspects of employment

The most important aspect of employment in Spondon's two largest industries was adequate control of the many dangers that were always present. In both the Power Station and Factory there was superheated steam up to about 600psi, electricity up to 33,000 volts as well as numerous toxic and flammable chemicals. Moving machinery was a constant hazard especially textile machinery where numerous wheels, pulleys and bobbins rotated at high speed. It was important that employees reporting for work each day felt they were part of a team where dangers were recognised and adequately controlled.

A number of records and log books from the year 1918 have been retrieved and show the thoroughness with which operations were carried out. Numerous activities are described in detail and they all carry dates, times and signatures. There is a record that during the 1930s Spondon had the sixth largest private power system in Britain and was amongst the first to adopt Live Line Testing as standard procedure to ensure that High Voltage connections were safe to touch before work commenced.

Legislation also played a part and electricity was governed by the Electricity (Factories Act) Special Regulations of 1908 and 1944, and chemicals by the Chemical Works Regulations of 1922. These were rigidly adhered to and when British Standard (BS) 227 appeared in 1929 for the control of electricity in hazardous atmospheres, Celanese immediately implemented it wherever acetone was used. UK power stations always had a record of self-sufficiency and Spondon was no exception. Outside services were only used for certain specialised requirements or major contracts; everything else was done 'in house'. A full range of skilled tradesmen was employed and nearly each one of them had a 'mate' who was paid at a slightly higher rate than a general labourer. So both Spondon's largest industries were seen to be very safety conscious by the standards of the day.

It is worth a few final notes on Health and Safety because over the decades attitudes and practices changed a good deal from 1918 when workers were expected to use their common sense and avoid contact with anything that would harm them, but slowly new ideas would evolve where Management was expected to take a more pro-active role in providing a safer environment.

The impetus in 1974 was two events that signalled major changes. These were the Flixborough disaster on Saturday 1 June and the introduction of the Health and Safety at Work Act.

What followed was a steady line of progress up to the present day, now a world away from that of 1918.

The other area in which they both excelled was training of all types. Apprenticeships in the Factory and Power Station ran for a full six years and involved supervised training in every part of the sites. Power Station apprentices spent their first six months in the Engineering Workshops of Derby Technical College and Factory craft apprentices spent their first year in the Company's own Training School. There was also a Girls' Typing School for selected female school leavers. All apprentices had to attend part-time day studies at the Technical College which comprised one full day and two evenings per week and their results were carefully watched. Most apprentices went on to remain with their respective employers and it was not uncommon for them to go on and complete up to 50 years of service.

Turning to other matters which directly affected employees' lives, the Power Station had a full meals service during normal working hours with separate dining rooms for works and staff. It was normal practice for night shift workers to bring their own refreshments and consume them on the job; but Alan Hunt tells how he and a colleague, when working a night shift, would walk through the station gate with a nod to the gateman then make their way to the Celanese Main General Canteen adjacent to the Main Avenue, where they would enjoy a break and a hot meal before returning. They were never challenged on their nocturnal journeys but it is unlikely that Celanese would have approved.

Christmas 1963

The Power Station had large shower rooms, kept immaculately clean, with soap and towels provided. There was also a staffed First Aid centre as well as a trained crew of part time fire fighters.

Not so well known was the fact that CEGB had built a golf course in the cooling towers area which was open to employees free of charge and to local residents on payment of a small fee.

Operating staff comprised every level from Shift Charge Engineer downwards and included Control Room staff, boiler and turbine operators as well as a full range of general labourers for coal, ash and other routine tasks. This latter group was the largest and numbered around 250 employees.

During normal working day hours there would have been trained engineers, laboratory staff, clerical staff, full time cleaners and supervisors.

British Celanese Fire Brigade with their Stanley fire engine around 1927

The strength of the brigade was increased on several occasions by the addition of more modern equipment, but this original unit was kept mainly as a museum piece, well into the 1980s. It was, however, always maintained in full working order, and members of the brigade claimed that it had a strength of jet far greater than many more modern vehicles.

The notice over the doorway in the background states 'Danger – no matches allowed inside this plant' and the notice alongside the smaller door forbids the carrying of lighters, matches, and quotes 'smoking strictly prohibited'. Due to the highly volatile nature of the products, smoking was not allowed anywhere on the Factory site except in the canteens. This rule was strictly enforced and anyone found infringing the rules received instant dismissal.

Bus Services

Altogether Spondon A employed between 400 and 500 people. It is worth noting that the Station also had its own bus service. This consisted of a Barton double decker which ran twice a day throughout the week between Cotmanhay and the Station, calling at Ilkeston, Kirk Hallam, Dale, Ockbrook and Borrowash, before proceeding to the Station via Power Station Lane.

Typical Power Station bus pictured at Huntingdon Street, Nottingham, operated by Barton. This one is a Leyland Titan.

Factory circular bus used to carry workers to and from the site at shift changeover times. This was a Trent service from Derby.

The Workers' Train 1920 to 1960

By 1918 plans were already in place for Spondon Station to have one additional platform on the south side of the down main line; the Directors must have had some confidence for the future of the Company. A copy of this drawing exists and there is also a photograph of the site works for the extra platforms, which were built for one purpose only, to provide transport for employees in the adjacent industries.

A heavy load for a Johnson 2-P Class 4-4-0 on the 5.37pm workers' train out of Spondon in April 1952. The first seven coaches were for Darley Dale and the last five for Crewe.

Services began in the early 1920s with what came to be known as the Workers' Train and there was a regular service throughout the week. Trains started from Darley Dale and called at every station on the way to Spondon, with Derby being an obvious major joining and leaving point. When it is recalled that at its peak the Factory and Power Station together employed around 20,000 people one can easily imagine just how busy these trains were. They arrived and departed at shift changeover times and had up to 12 coaches pulled by a Stanier or Fowler tank engine or a Johnson 4-4-0 tender engine. On arrival at Derby the rear five coaches were attached to the rear of the waiting Crewe train so that workers could continue their journey.

The Spondon to Derby section must have been intensely crowded with both seated and standing passengers. On the journey from Spondon the train would be full of workers who had been unable to smoke for a whole eight hours, so it is easy to imagine what the atmosphere must have been like with the windows closed on a dark day during the winter.

By today's standards these conditions would be totally unacceptable so it is not surprising that the Workers' Train did not survive beyond the age of steam and the end came around 1960 when the Diesel Multiple Units came into use on suburban routes. British Rail would never have allowed their smart new 'state of the art' trains to be filled with Factory workers fresh from an eight hour shift in a large chemical factory which had something of a local reputation for smells and industrial working conditions typical of the period.

Murder at the Power Station

At around 9am on the morning of Sunday 3 January 1943 the Superintendent of Spondon A was descending the stairs at the south east corner of the turbine hall after spending the night on the station roof doing his spell of fire watching. This was an obligatory requirement for most men who were not serving in uniform and the Superintendent, although holding the most senior post on site, had to carry out this duty on a regular basis.

The Superintendent, Ernest Hezekiah Lynam, entered his room; as he left he was confronted by the duty Shift Charge Engineer, Cecil Henry Dawson, who produced a loaded revolver and shot Mr Lynam in the neck before firing another shot into his own head. When police searched the bodies it was found that Dawson was also carrying two loaded automatic pistols, one of which was a Mauser.

Mr Lynam was a married man aged 46 and his address was given as Just Holm, High Lane East, West Hallam; it was reported that he had held his current post for some years. Dawson was a single man aged 41 who lived in Drummond Street, Ilkeston.

Some additional information has been provided by Alan Hunt of Spondon who was an apprentice electrician at the Station between 1958 and 1964. He relates how a Senior Apprentice Electrician, Geoff Coopey, showed him at an early stage in his training what were described as bullet holes in the brick wall of the stair well where the killing took place. Alan recalls seeing several holes.

It seems likely that when Mr Lynam saw Dawson pointing a gun at him his immediate reaction would be one of flight and he would attempt a rapid departure from the scene with Dawson in pursuit. This would explain the possibility of up to five shots before the fatal neck wound because the most obvious target would have been the large area of the chest. It seems very likely that Dawson was firing at a moving target and from that point on there would have been a high degree of uncertainty as to where the fatal shot would strike.

No motives have been uncovered to explain the reason for this crime.

4 Spondon H Power Station 1959 to 2017

This was a Combined Heat and Power Station (CHP) and its steam turbines were arranged to supply exhaust steam at a set pressure for heating and process purposes, all of which went into the Factory site.

Spondon H when new, with Spondon A to the left

This is the point at which the highest saving is made because in conventional schemes the exhaust steam has to be condensed back into water which causes a large loss of energy; CHP schemes do not have cooling towers but on the downside they do require a high-volume water-softening plant to provide lime free water to feed the boilers.

Because the energy match is never perfect there has to be a way of using the surplus of either steam or power to a financial advantage. This is normally done by connecting the generators to the National Grid so that surplus electricity may be disposed of and a payment will be made for the exported power, but at something less than the full commercial rate.

The Spondon Factory site had always been an ideal choice for a CHP scheme and not surprisingly, both CEGB and Celanese engineers had identified the site a number of years previously. David Howard, who went on to become the Manager of H, has recalled seeing an item of correspondence in 86 Building Boiler House offices which referred to an application to the Ministry of Supply in 1946 to build a high pressure boiler plant. It seems that the Ministry turned down the application because it called for high quality steel at a time when virtually all materials were in very short supply and urgently needed for post war reconstruction.

By the early 1950s times had moved on and CEGB knew that the days for Spondon A were numbered. Likewise, British Celanese had four boilers dating back to 1917, so all of the facts fitted for the planning of a CHP scheme.

Exactly what happened next is uncertain but it seems likely that CEGB put some ideas to Celanese, who then prepared plans for the design, financing and building of a CHP scheme to be erected on their site. The plan was that they would operate it, put all of its electricity into the National Grid and the steam flow through the turbines would match the needs of the Factory, and Celanese would pay for it.

There is firm evidence of their mutual interest in this project. We have a copy of the British Electricity Authority Drawing Number EM4725 dated 8 July 1954. This shows the Spondon A layout and a full outline of Spondon H, almost exactly as it was completed six years later. The drawing carries OS Number 19384, (Outside Print) proving that it was received by British Celanese. In due course a firm proposal was put forward and a twenty year contract agreed, paving the way for site work to commence July 1956.

Spondon H Firing Floor

The final boiler and turbo generator were commissioned in October 1960 and the Station was officially opened on 21 April 1961 by Sir Robertson King, Chairman of the Electricity Council.

CEGB went on to supply steam to the Factory at the rate of about 179 tons/hour for the next twenty years but for most of this time the Factory boilers in 86 Building were still required to top this up to about 268 tons/hour.

30 inch steam main to Factory being installed in the late 1950s

British Celanese events which paved the way for Spondon H

Two new boilers each rated at 15.6 tons/hour were added to the ten existing boilers in 86 Building in 1935, and the four original boilers in 13 Building and 23B Building. were retained in service. David Howard believes that two extra boilers were relocated into 86 Building making 14 in all. Peter Heath remembers seeing the 23B boilers in service during the early 1950s so it seems most likely that these were the two which were reinstalled into 86 Building.

In 1948 a total of ten Lancashire boilers each rated at between 3.6 and 4.5 tons/hour were installed in 86A Building. These were able to burn a variety of liquid and gaseous fuels, much of which was either surplus or waste from other parts of the site. Some of them could also burn heavy fuel oil so Celanese was able to make the most efficient choices in its disposal of energy sources already on site.

The Directors' optimism in the 86 Building investments was shown to be justified by ensuing events, as shown below, and all of these seven major investments were to follow in rapid succession.

- 79 Building - Filter Cloth Recovery indicating progress in Dope Preparation.
- 80 Building - A large new toilet and cloakroom block indicating more jobs.
- 81 Building - Dope Mixing Rooms.
- 82 Building - Acetone Recovery, a large new plant which succeeded 56AR.

- 83 Building - By far the largest Spinning Room so far, to be known as 5SR.
- 86 Building – Main Boiler House at junction of Nottingham Road and Main Avenue
- 88 Building – Using the progress made in weaving and dyeing techniques.
- 94 Building - Completed in 1935; this was said to be the largest weaving shed in the world, 1,000 feet in length and 220 feet wide with the northern half of it having either three or four floors.

By 1935 Celanese had come of age, was in the big league and had become profitable. At this time, the Factory steaming capacity was 156 tons/hour and 1948 saw this rise to about 196 tons/hour. So by 1948 there were 22 boilers at Spondon A Power Station and 26 in the Factory, a total of 48 in all and each had its own chimney producing either steam or electricity, but never the two together.

By 1946 the Factory's high voltage distribution system had remained unchanged since 1917 and had become dangerously outdated. It was exposed to a huge fault level from a now fully grown Spondon A against which it was quite defenceless and had long since been overtaken by the relevant BS116 of 1937. Special procedures had to be adopted to limit the dangers to which switching engineers were exposed and something had to be done. It was decided to build the Boundary Substation, Building 433, on the north side of 29 Building and supply it directly with two 33,000 volt circuits from Spondon A. This was a huge 16 panel switchboard divided into four separate sections, each one housed within its own fireproof compartment and it provided eight new outgoing circuits to feed power to the Factory.

The main contract was undertaken by Balfour Beatty Ltd and they designed, built and equipped it with two 33,000/6,600 volt 15MW transformers connected to a very large oil break switchboard whose circuit breakers had been independently tested to break a fault power of 350 million volt amps.

The main safety concern had been allayed but there still remained most of the site substation circuit breakers which dated back to the time of the First World War and virtually all of these were inadequate for their duty.

The Boundary Substation entered service in 1948 and between 1950 and 1955 all of the site High Voltage system comprising the five major Substations A, B, C, D and E were brought up to BS116 standards and certified to break 250 million volt amps of fault power.

Boundary substation, middle of picture, erected 1948. Photograph dated 1990s. To the rear are the Factory effluent tanks and at the front is the major power factor correction scheme of 1955.

As a final 'belt and braces' measure there was a steam line from Spondon A to Spondon H rated at 45 tons/hour at 375psi but there is no evidence that this was ever used.

The layout of Station H
Each boiler had a large control panel equipped with a comprehensive range of meters and indicating instruments covering all of the main operating conditions and this was for the use of the firing floor operators. They could also set the chain grate speed and the depth of the fire bed.

The main control room for the Station was situated on the east side of the Turbine Hall and raised slightly above it. The main control panel displayed the main steam and water conditions with instruments at each of the appropriate positions. To the left was the diagram of the main high voltage system and on the right hand side was a similar panel for the three generators.

**Control Room showing mimic diagram
and electric control panel 1960**

Adjacent to the south wall was the main control panel for the 33,000 volt switchboard which provided connections to the East Midlands Electricity Board power system.

Next to the Control Room was the Alarm Panel which provided audible and visual alarms for almost every aspect of the operation of the station.

Without going into details, the main power circuitry of Spondon H was large and complex with a number of mathematical aspects. It is sufficient to say that a variety of voltages and power levels were involved, including 33,000 volts, 11,000 volts, 3,300 volts, and standard 415/240 volts for ordinary power and lighting.

Generators 8 and 9 were connected directly to the 11,000 volt switchboard but Number 10 went onto the 33,000 volt board via a 12.5MW transformer. This was to prevent the 11,000 volt board from being overwhelmed by short circuit currents which would flow in the event of a serious electrical fault.

The same reasoning applied to the two 33,000 volt EMEB supplies connecting the Station to the National Grid because the Station's board had only 50 percent of the capacity of the EMEB board in terms of its ability to carry excess currents. This problem was overcome by installing one set of air insulated reactors in series with each of the two supplies so that the fault current level was about 50 percent of the EMEB level. Each reactor produced a voltage drop of 15 percent at full load, sufficient to relieve the Spondon board of excess duty in the event of a fault.

The 33,000 volt switchboard had fourteen separate circuits and six of these were named CAL1 to CAL6, the abbreviation CAL referring to Courtaulds Acetate Ltd. They fed the Factory site and each one supplied a 33,000 volt to 6,600 volt transformer located at the Boundary Substation. Four were rated at 15MW each and the other two at 20MW. They operated as three pairs so that the Factory had three separate 6,600 volt distribution networks.

33,000 volt control panel 1960

Commercial aspects of Spondon H
The station was never intended as a serious producer of electricity. Its output was almost miniscule when compared to the other CEGB stations dotted along the Trent Valley. Its purpose was the sale of heat energy in the form of steam and that to a single tied customer. To that end a 20 year contract was drawn up between the two parties in order to ensure their future security.

One of the main features in the contract was known as the Escalation Clause. This was written in to ensure that CEGB would not be unfairly penalised by fluctuations in the price of fuel and the Clause was applied at each month end so that minor adjustments could be made to the bill. The arrangement worked well until the early 1970s, when the UK entered a period of hyperinflation and at that point the Contract failed to produce an equitable adjustment, it being heavily biased towards the customer's interests. CEGB appealed to the Factory Directors to recognise their plight and accept an increase which would at least enable them to balance their books but the Factory was adamant in their refusal and said that the Agreement should stand.

Fortunately for CEGB there was a minor clause in the agreement which allowed the President of the Institution of Civil Engineers to appoint an adjudicator to hear the case if a serious dispute arose between the two parties and so the matter went to an official hearing. Within a day or two the Adjudicator found wholly in favour of CEGB and a suitable variation was made to the Agreement.

Very shortly afterwards the Factory Chief Executive left the site, never to return, and there was a general view that he had paid the price for his obduracy in failing to seek a compromise with CEGB.

This was the period when activity in the Factory was at its highest level in the Company's history. Peter Heath remembers a summer's afternoon in 1974 shortly after the arrival of the 2pm shift when the total load indicator in his office peaked at 43.4MW, its highest ever level.

For several years previous to this, the Factory had undergone an average annual load increase of about eight percent per annum and it was obvious that the supply system was at its limit for safe and stable operation. In particular, Substation B on Fleet Street was handling 1,200 amps at 6,600 volts for 24 hours a day on four feeders arranged in two parallel pairs. It was known that the loss of just one cable would lead to cascade tripping as the load moved over to the remaining three cables and they would automatically drop out of service due successive overloading. It happened, but not to a cable, a whole switch cubicle burnt out instead and the Factory was shutdown for 24 hours while essential repairs were carried out.

At last, minds of senior management were focussed on the need for some serious expenditure and within a day or so the money was found to completely re-supply the area bounded by Eastern Avenue, Main Avenue and Filey Road as well as the 148 Building areas. The scheme included eight new 6,600 volt feeders, two 20MW transformers, two more 33,000 volt supplies from the Power Station and all the required new switchgear. The whole project was handled 'in house' from the Factory Substation's Office including procurement, erection and final testing. The final cost was about £250,000 and the works took fifteen months to complete.
At this time the Factory steam load stood at around 268 tons/hour, two thirds of which came from the Spondon H and the rest from 86 Building boiler house.

Winding Down
Activity continued at a high level on the site until the late 1970s and then the first closure announcement came along regarding Bonshawe (Fabric Print in 88 Building). It was a shock and others would follow such as Derwent Dyers, Bon Ton, Steel Cords, National Plastics and then the really big one with Alcohols, previously known as 1200 Plant. Eventually, Furzebrook (18 Building) would close so that by the early 1980s the

Factory site steam and electricity loads had fallen dramatically, maybe by as much as 40 percent, so there was no longer a need to produce steam on site and it could all come from Spondon H.

By 1982 CEGB had been operating Spondon H for 22 years and they must have decided that it was time to let it go, so it was offered to the Factory and bought by them for a figure thought to be about one million pounds.

This resulted in major changes for the Station. Firstly, the number of employees fell from 200 to around 100; secondly, the electricity produced could now be credited directly to the account of the Factory. The Company were now able to look for ways of optimising the production of electricity against steam because this was where some real money could be saved. They began by raising the degree of superheat imparted to the high pressure steam as it left the boilers, thereby giving it more energy in order to get more work out of the original AEI turbines, installed by CEGB when the station was built.

Next, in 1984/85, they looked at the AEI designs which by this time were probably about 30 years old and found that more modern designs were available having a much higher efficiency. Consequently, two new turbines were purchased from AEG of Germany and these ran at the very high speed of around 13,150 rpm which was then reduced to the standard 3,000 rpm through epicyclic gear boxes made by W H Allen of Bedford.

AEG 10MW turbo generator Number 9 1987

By 1987 Numbers 9 and 10 turbines had been renewed and David Howard recalls that the Company then benefited by about an extra 4 MW of power, with a payback time on the capital cost of eighteen months to two years.

The mechanical efficiency of the new turbines was greater than the old ones but the steam leaving them also had reduced superheat at 232°C instead of the 271°C of the

AEI machines. De-superheating then reduced the final temperature to a little over the natural temperature of steam at 170psi which is about 190°C.

By 1987, with all the closures that had taken place, the steam load was down to a level that could just about be supported by Spondon H alone. The time had come to let go of 86 Building with its long history of the Babcock Boilers dating back to 1927. Before demolition could start it was necessary to take some steps to ensure that there was full security for the Factory's steam supply. This was achieved by bringing in two oil fired Robey boilers that were no longer needed at Courtaulds Grimsby Works and installing them on the ground floor of the Power Station at the east end of the Turbine Hall in1988. Their output pressure was 170psi and they were used whenever it was required to augment the output of Spondon H in order to meet the Factory's steam demand.

Efforts to increase efficiency continued and over a period of time the labour force was successively reduced from about 100 to 60 but with the added bonus that the Shift Team had also taken on the extra responsibility of overseeing the Factory's cooling water supplies and effluent services.

Much work was also done in order to diminish the importance of coal as the Company's only energy source. To this end numerous trials were carried out with various alternative fuels that could be mixed with the coal before it was burnt on the chain grate stokers. Amongst the fuels tested were chopped redundant vehicle tyres, pelletised refuse, wood chippings and a substance known as Petcoke which was obtained from oil refineries. This latter fuel was a semi-solid sticky black waste product that burned with a ferocious heat, as did the chopped tyres. It was said that burned alone they were capable of melting the cast iron grates or shortening the life of the grates to the point where the use of these fuels alone was not a practical proposition. All of these experiments were worth pursuing and some successes were achieved, but it could not be claimed that they led to any major long term savings.

There was one other alternative fuel named Orcol that was used on a regular basis for the oil fired boilers and that was reclaimed oil, mainly from commercial garages. It was delivered in large road tankers and had a clean and innocuous appearance but analysis of the flue gases and boiler residues revealed that it contained various trace metals such as lead and vanadium. Eventually its use was terminated and perhaps that was the reason. Certainly it is difficult to imagine that such a fuel could be burned today and the flue gases be discharged into the atmosphere. Unlike the other alternative fuels it was easy to burn, was relatively cheap and for a time was used in quite large quantities.

By the end of the decade Spondon H had been operated by the Company for about ten years and had evolved into a highly efficient supplier of steam and electricity to the site. It comfortably supplied all of the steam as well as most, if not all of the electricity, and at times there was a surplus of electricity which was simply returned to the grid. CEGB made a small payment for this energy but it was only just enough to cover the production costs and in any case it was simply 'spill over' power as a result of the steam demand.

During the same decade the Factory had also seen a number of closures, many of which were quite small, but three major changes had taken place affecting steam and power demands.

The first of these was the Alcohols Plant which was the starting point for making acetic acid and much could be written on this subject, about its rise and ultimate fall but this is not the time or place for it. It is a story on its own with a number of interesting aspects and ought to be recorded. Alcohols Plant was a major user of steam and power with a connected load of somewhere around 5MW.

The second one was Furzebrook Knitting and Dyeing in 18 Building. They were major users of steam and had a medium sized power demand.

The third one was Celon built in 1965; it occupied the whole of 52 Building and was the Company's Nylon Plant. It had a fair sized steam load but was a major electricity consumer with a load approaching something like 5MW. A major contributory factor in the closure of Celon was the huge explosion at the Nypro Plant close to the village of Flixborough on 1 June 1974. This plant made Caprolactam, the basic raw material for the manufacture of Nylon fibres produced by Celon and its loss made commercial life much harder for the Company. It wasn't the only reason for the closure, the other being that demand for the type of Nylon made by Celon was falling rapidly so the eventual end of the Company was inevitable.

The net result of all the closures up to 1990 was to place a total load on Spondon H of about 178 tons of steam per hour and no more than 25MW for electricity, still a sizeable load to supply all of the core activities that remained on the site and were still profitable.

One more closure would take place in 1992 and that was the transfer of the Novaceta activities to Italy. This ended about 70 years of textile yarn production at Spondon. The Novaceta spinning machines were then converted to produce cigarette filter tow, by then the Company's main source of income.

Commercial activities continued on site and in 1995 Spondon's fourth and final power station was completed. It had been designed from the outset to supply high pressure steam directly to the turbines at Spondon H so that from then on all coal burning at Spondon ceased and Boiler Numbers 23, 24, 25 and 26 fell cold and silent.

It was the end of an era, of a time when the two Spondon sites had had about 48 coal fired boilers between them and supplied about five percent of Britain's electrical power needs.

5 Spondon's last Power Station and Derwent Cogen - 1995 to 2012

By the early 1990s Spondon H boilers were approaching an age of about 35 years and there was a general recognition that their end was in sight.

David Howard recalls that at this time he began work with George Philipson, Director of Engineering at Spondon, on some form of Combined Cycle Gas Turbine (CCGT) plant. This would produce both high-pressure steam for the Factory and export a sizeable amount of electrical energy directly into the National Grid at 132,000 volts. Initially it was thought that a local solution might be at hand by using Rolls-Royce industrial gas turbines as a suitable starting point. Some initial progress was made along this route but then it was discovered that an American company known as Mission Energy were already in the UK and it wasn't long before talks had begun with them to look for some alternative way forward.

From this point on events moved quickly and Mission put forward a CCGT scheme that would fulfil all of the requirements already envisaged in terms of both steam and electricity. The plan was for a dual fuel station using natural gas as a first choice but also able to operate on diesel oil at times when the gas pressure was too low or gas might not be available due to national needs.

A flexible approach was arrived at using gas turbines exhausting through steam generators which could supply steam at about 65 bar (950psi) either into the Factory or directly into a 54MW standard condensing steam turbine. By this means the Station would be able operate on a standalone basis even if the Factory ceased to exist and could export something like 210MW of power into the National Grid.

Site works began in 1992 with the main civil contract going to Trafalgar House and the technical aspects under the control of Ebasco, a New York company, later to be part of the Raytheon group. The station was of modular design with all of the main components housed in separate weather-proof enclosures. It was said to be a virtual copy of a similar plant in the Nevada Desert, sited not far from Las Vegas. When the station was in service, it was not unknown for engineers to comment that the climate of Las Vegas was not the same as that of a Spondon winter, where working outdoors could be something of a hardship.

Here are some details of the Station's main components.

Gas Turbines
There were four, each with an output of 40MW at a turbine speed of 5,150 rpm and were geared down through plain spur gear boxes to a speed of 3,000 rpm to produce power at the standard frequency of 50 cycles per second. The turbines made by General Electric of America were classed as Frame 6-6001B and could be fired by either natural gas or diesel oil, but the first choice of fuel was always gas.

Each turbine had ten combustion chambers placed at the end of its compressor stage and the gas supply was of a sufficiently high pressure not to require any assistance in entering the chamber.

Exhaust temperature was 640° C and, as is the case with gas turbines, this hot gas stream was very oxygen rich, so it was a simple matter to inject additional fuel into it thereby raising its temperature still further. This was achieved by having six gas burner injection points attached to each duct carrying the heated gas stream into the four boilers. They were made in Derby by International Combustion Ltd. Known as Heat Recovery Steam Generators, each one had its own chimney.

The turbines were operated as two pairs, A, B, C and D, referred to by the staff as Alpha, Bravo, Charlie and Delta The hot gas stream from each machine was fed through a Heat Recovery Steam Generator to supply steam at 950psi to the Factory.

Alpha and Bravo gas turbine generators 1995, photographed in 2016

Steam turbines
There was one, rated at 54MW, fed from gas turbines Charlie and Delta producing steam via the Heat Recovery Steam Generators. It had always been envisaged that, if ever the Factory were to close, a second steam turbine could be installed. This would produce a balanced system without the need to export any high pressure steam, but it seems that the closure came too late for any more capital expenditure.

Fuel
The gas came in at a pressure typically between 22 and 30 bar (323 and 441psi) via an eighteen inch diameter steel pipe so it could be fed directly into the gas turbines without the need for further pressurisation. The maximum delivery pressure was said to be 35 bar but this would have been very unusual and for direct feeding the minimum pressure was said to be 17 bar, but 20 bar was then regarded as the point where a changeover to diesel oil was made. This did arise on a few occasions up until 1998 because there was always a clause in the contract that the supply could be cut off with little warning at times when the national need was greater. After 1998 the gas supply was reinforced and if any further interruptions did occur they were very few and far between.

Gas connection point

Diesel oil was held in three separate steel tanks with a total storage capacity of 12 million litres.

Diesel tanks

Auxiliary boiler
This was a stand-alone boiler with a nominal capacity of 20 tons/hour at 985psi and was normally gas fired but it could also burn oil. On the few occasions when it was used it was said that its actual delivery was closer to 13.4 tons/hour than the nominal rating.

Cooling water
This was drawn directly from the River Derwent.

Boiler feed water
Extreme purity was a requirement and it had begun its life at Spondon H where it had already been used for some minor cooling purposes, so it arrived at the CCGT Station at a slightly elevated temperature. It was supplied without charge at a rate of 300 cubic metres per hour and then went on through an elaborate and complex purification process from the end of which it emerged as virtually pure water. This degree of purity was essential as it went into the Heat Recovery Steam Generators and none of it was returned from the Factory after being used for power generation and process heating.

Inlet air for gas turbines
This has to be clean and pure fresh air and it is essential that there are no foreign bodies or particles in it. To this end the air is drawn through a large filter system before it enters the first stage of compression in the turbine. The quantity of air required is very large but only a fraction of the oxygen content within it is needed for combustion purposes; the remainder passes out in the exhaust stream at about 640°C.

Staffing levels
A total of 41 people were employed consisting of five shifts of three plus a permanent staff which included seven technicians.

Cold start
The time taken from a cold start to full power was only 45 minutes. An initial warm up was followed by a loading increment of 2MW per minute after the first 20 minutes.

Power generation
The station had a total of five generators; four gas turbine driven, rated at 40MW each, plus one steam turbine driven machine rated at 54MW; a total gross output of 214MW.

Unit transformers convert the power from the generators so that it could be sent to the National Grid and the Station had three of these, UT1, UT2 and UT3 rated respectively at 90, 110 and 45MW.

Generators Alpha and Bravo used UT1, Charlie and the steam turbine machine used UT2 and Delta exported via UT3. All of the Units exported at 132,000 volts directly onto the National Grid.

Derwent Cogen (gas turbine station) was originally called Derwent Co-Generation Limited. The name arose from the dual purpose of the station which was to run both steam and gas turbines together as well as supplying high pressure steam as required to the Factory for yet more electricity production with a final run off of the last bit of remaining energy in the form of steam for processing purposes. The whole set up was just about the ultimate in thermal efficiency; almost nothing was wasted.

Final Closure
The station ran very successfully from March 1995 to November 2012 when the Celanese Factory closed its gates at the end of its last shift.

The Steam Contract expired in October 2010 but went on to have another two separate one-year extensions taking it to October 2012, just prior to the factory closure in November of that year.

After the Factory closure there was still the need for an orderly shutdown of the works steam system as well as supplying Clarifoil with steam until its own new boiler was in service, so for a few weeks high pressure steam continued to be sent into the Factory. A side effect of this was to produce about 28MW of surplus electrical power and this was simply discharged into the Grid.

The end of steam and power generation came at 2pm on the 12 December 2012, 95 years of unbroken continuous production of electricity and steam at Spondon, but it would be a further 4 years before the Spondon sites were finally made dead at 2.30pm on 16 February 2017. During this remaining period power came in at 33,000 volts from the National Grid so that lights could be kept on and Clarifoil supplied with power at the normal level of 440 volts.

Shortly after generation ended the 132,000 volt Grid connection was removed and a section of the high pressure gas supply main was taken out, so the CCGT station now stands isolated and awaiting a prospective purchaser for either the whole package or failing that the sale of individual major components. If no buyers come forward, then in the not too distant future it will almost certainly be cut up for scrap and nothing will remain.

Spondon's power and steam story came to an end in 2017. As a final observation it is worth noting that Spondon A required a staff of about 500 to produce 170MW from coal and the final station only needed 41 to produce 210MW from gas as well as providing the Factory with all of its high pressure steam needs.

Both Spondon and Derby Power Station sites will continue as major distribution centres into the indefinite future and distribute power to a wide area of the Midlands.

Acknowledgements

Spondon Archive thanks the many people who have helped with this book, particularly:

Anthony Rawlinson and Kevin Brocklesby for photography

Ashover Light Railway Society for photograph of electric loco

Bert Loomes, Water and Effluent Services Engineer, 1976 to 1998

British Celanese Management for access to Site, Maps and Drawings

David Hayes for editing original text

David Howard, Manager of Spondon H Power Station, 1983 to 1994

David Wimshurst for access and help with the Gas Turbine Station

Derby Telegraph for pictures from Bygones

Derek Hathaway for scanning photographs and technical help

George Philipson, Engineering Director, 1976 to 1992

Jack Hartley for pictures of Steam Main Contract

Oliver Cotton, Maintenance Manager, Gas Turbine Station, 2007 to 2010

Steve Allen for cover design and help with photographs

WW Winter Ltd. for pictures of Derby Electric Light Works

Many of the photographs are from Alan Hunt's collection, and in some cases their original ownership is unknown. Our apologies if we have inadvertently breached copyright. Please advise us if you own the copyright and we will give a correct attribution in future editions.